IMAGES
of America

GLENVIEW
NAVAL AIR STATION

A Navy Flyer's Creed

I am a United States Navy flyer.

My countrymen built the best airplane in the world and entrusted it to me. They trained me to fly it. I will use it to the absolute limit of my power.

With my fellow pilots, air crews, and deck crews, my plane and I will do anything necessary to carry out our tremendous responsibilities. I will always remember we are part of an unbeatable combat team— the United States Navy.

When the going is fast and rough, I will not falter. I will be uncompromising in every blow I strike. I will be humble in victory. I am a United States Navy flyer. I have dedicated myself to my country, with its many millions of all races, colors, and creeds. They and their way of life are worthy of my greatest protective effort.

I ask the help of God in making that effort great enough.

Seen here is the navy flyer's creed.

IMAGES
of America

GLENVIEW
NAVAL AIR STATION

Beverly Roberts Dawson

ARCADIA
PUBLISHING

Published by Arcadia Publishing
Charleston SC, Chicago IL, Portsmouth NH, San Francisco CA

Printed in the United States of America

Library of Congress Catalog Card Number: 2006931411

For all general information contact Arcadia Publishing at:
Telephone 843-853-2070
Fax 843-853-0044
E-mail sales@arcadiapublishing.com
For customer service and orders:
Toll-Free 1-888-313-2665

Visit us on the Internet at www.arcadiapublishing.com

*This book is dedicated to the men and women who served at Naval Air
Station Glenview, 1937–1995.*

CONTENTS

ACKNOWLEDGMENTS

It has been one of the greatest privileges of my life to make the acquaintance of many of the men and women who served at Naval Air Station Glenview (NASG). I am indebted to Bill Christensen, one of the original station keepers at Glenview, who introduced me to the story of the base and provided expert assessment of the manuscript. Eric Lundahl, A. C. "Ace" Realie, and Charles Downey shared their recollections and photographs. To Leslie Coker, editor of *The Last Salute*, goes a huge vote of thanks for facilitating preservation of volumes of NASG photographs and written materials, which were of immeasurable help in researching this book. NASG veterans Jack Witten, Ted Koston, Bill Schoewe (and members of the Glenview Survivors Association), Bill McConnell, Mary Fenoglio, Frederick C. Durant, and John Larson (along with the men and women of VP-90) contributed a wealth of information. Therese Gonzales, curator of the Great Lakes Naval Museum, provided rare vintage photographs. NASG daughters Debbie Poynter Gust and Linda Schram Welsh shared family photographs and anecdotes. Glenview Area Historical Society volunteers Marge Marcquenski, Dorothy Murphy, and Virginia Peterson, along with Irving Danneil and Bob Coffin of the Glenview Hangar One Foundation, devoted hours of assistance. Dick Zander and Don Long shared their recollections of growing up in Glenview before and during World War II.

I would like to thank Arcadia editor Jeff Ruetsche for recognizing that this book needed to be written—and then allowing me to do it. The technical aspects of putting together a book of this nature are formidable. A tremendous vote of thanks goes to my husband, Dr. Bill Dawson, for spending hours scanning and archiving the photographs herein. I am indebted to my daughter, Susan Dawson-O'Brien, journalist par excellence, who provided encouragement and professional advice. Perhaps I owe the greatest vote of thanks to my grandson Evan James O'Brien, who is an ongoing source of inspiration. It is my hope that he and his generation will carry forward the superior performance demonstrated by those who served at NASG.

— Beverly Roberts Dawson

INTRODUCTION

How can one properly present the story of Naval Air Station Glenview in a few short pages? In truth, it is not possible to cover every facet of this fascinating installation; perhaps the best approach is to begin at the beginning.

The first immigrants arrived in Illinois to the region now called Glenview in the early 1830s. Mostly Europeans of English and German origin, they were homesteaders who worked hard to buy land that the government offered for $1.25 per acre. As time went by, many of the landowners sold off parcels, resulting in smaller farms with fewer acres, growing vegetables and flowers. In early 1929, several farmers who owned the property sold their land to the Curtiss Flying Service; work to build Curtiss Reynolds Wright Airport (variously known as Curtiss Reynolds, Curtiss Wright, Curtiss Chicago—or simply Curtiss) began almost immediately. The airport had barely opened when the New York Stock Market crashed on October 29, 1929. The airfield stayed afloat during the Depression years largely due to the entertainment side of the business. While people may not have had enough money to learn to fly, many could afford the price of a ticket to watch the most popular form of entertainment in the land—air shows.

Although there had been a naval aviation unit in place at Great Lakes for several years, the progress of aircraft design and technology made it necessary to relocate. Since Curtiss Wright Airport was a modern facility, it seemed an ideal place to expand the navy's flight training program. Those who facilitated the transition from Great Lakes to Glenview to establish United States Naval Reserve Aviation Base (NRAB) Chicago were a small but remarkable contingent. Excellence was the minimum requirement at NRAB Chicago. Jack Witten, who served at the base in the late 1930s, recalls, "a cadre of talent . . . permitted NAS Glenview to grow from the Naval Air Reserve Base . . . to a full blown Air Station. To accomplish this growth and still find time and talent to train the embryo pilots in the rudiments of line safety, so they didn't lose their arms or head in a propeller, and so they wouldn't lose themselves in the hegira of flight which would take them from Glenview . . . ultimately to the oceans and seas of the world."

After December 7, 1941, the pace at the base accelerated. As the United States expanded its war efforts, it became obvious that the number of pilots required would be unprecedented. The need was particularly urgent in the Pacific, where planes and carriers were to play a major role in the war with the Japanese. Comdr. Richard C. Whitehead conceived the idea of refitting commercial vessels to serve as training carriers on the safe inland waters of Lake Michigan. It was a brilliant concept for three major reasons: German U-boats prowled the waters off the Atlantic Coast of the U.S., making any carrier training operations very dangerous; Japanese submarines were often spotted off the Pacific Coast of the country, so prospects of maintaining an effective program off the West Coast was deemed very chancy; last, but certainly not least, America needed every carrier in its inventory to be deployed with the fleet. From that concept grew the unique Carrier Qualification Training Unit's "Corn Belt Fleet"—the USS *Sable* and USS *Wolverine*. It would hardly be possible to overstate the pivotal role NASG played in the winning of World War II.

After peace was declared in both Europe and the Pacific, "demobilization" was a hot topic in the base newspaper, the *Exhaust*. The strength of the navy at the time was more than 3 million; it was projected that half the personnel would be discharged in a six-month period. Sailors and WAVES (Women Accepted for Voluntary Emergency Service) expressed happiness at the thought of returning to civilian life but lamented that shipmates would be sadly missed. The great number of marriages taking place at the base chapel indicated the strength of bonds that had been forged among many of those serving at NASG. Among those being demobilized in 1946 was Lt. Comdr. Gerald Ford. Having been stationed at Glenview for the final six months of his World War II tour of duty, he was discharged from the navy. Reporting for duty in 1942, he served as a military training and fitness training officer at navy pre-flight school in North Carolina and then was assigned to the USS *Monterey*. He became the 38th president of the United States 28 years later.

Following the Japanese surrender on August 15, 1945, attention rapidly shifted to transitioning huge numbers of veterans back to civilian life and disposing of equipment and stockpiled materials no longer required for the war effort. The *Sable* and *Wolverine* were on the "surplus" list, and both were ultimately sold for scrap. It seemed an unceremonious end for two ships that had contributed so much to allied victory in the Pacific war.

Only five years after World War II was over, the Korean War began on June 25, 1950. Thirty thousand members of the Naval Air Reserve were called upon to augment active navy personnel. Navy and marine squadrons from NASG were deployed and served with distinction in Korea.

The postwar years were not all military business; the 1950s were busy times for public outreach at NASG. The base hosted national model airplane meets, and its personnel participated in Chicago metro-area sporting events. The air station baseball team played a series with the Chicago Police Department, and the navy boxing team participated in a show at the Chicago Armory. The Chicago premiere of the film *South Pacific* was held at the base. A navy color guard was formed and regularly appeared at area parades and picnics.

The cold war era continued with numerous changes at NASG. The base received a new tenant when the Seabees consolidated forces there in 1963. Reservists formerly based in Chicago and surrounding suburbs, along with units from Indiana and Michigan, came together to form Mobile Construction Battalion 27. During the time the Seabees were aboard NASG, they often worked with base public works personnel on projects to benefit the entire air station.

The 1968 Democratic Convention was held in Chicago on August 26–28. It was a time of tremendous social unrest in the country. Federal troops arrived at NASG on August 25. Requested by the State of Illinois and Chicago mayor Richard J. Daley, their mission was to stand by in case they were needed to quell disturbances in the city. Ninety-five planeloads—1,934 troops—from Fort Sill, Oklahoma, arrived and were bivouacked on base grounds. The convention was one of the most tumultuous in American history.

Not all tragedies were aircraft-related during the lifespan of NASG. In January 1967, a raging house fire claimed the life of Adm. Richard L. Fowler. At that time, the admiral was chief of naval air training at NASG. His wife and children sustained burns but survived.

The base expanded public outreach programs in a variety of ways. NASG opened its gates to dog shows, model airplane meets, rocketry clubs, and local civic events. The navy sponsored Sea Cadet groups and other youth-focused activities. A flying club and the NAS Sportsman's Club (featuring hunting, fishing, archery, and a pistol range) were open to base personnel and invited guests.

Based on water availability and other issues, the decision was made in 1971 to officially annex NASG to the Village of Glenview. At the time, the air station's census was 1,100 active duty navy, marine, and Coast Guard personnel, 600 civilian employees, and 3,000 navy, marine, and army drilling reservists.

During the Vietnam War, the Military Airlift Command C-9 Skytrains flew medical evacuation missions to Glenview, bringing wounded personnel for treatment at Great Lakes Naval Hospital.

On March 3, 1973, the first Vietnam prisoners of war arrived at NASG on their way to Great Lakes. Glenview rolled out the red carpet for the men and their families.

The decade of the 1980s saw many changes and milestones at NASG. A census taken in 1980 counted 100 aircraft and 1,300 full-time military and civilian personnel and 30 reserve units. The main (west) gate area was renovated with a new sentry booth and parking facilities. A quarter million dollars was allocated to refurbish barracks. New family housing was constructed, as was a new mobile home park. The Seabees and Glenview Public Works combined forces to build playgrounds for the children of NASG personnel. Squadron VP-60 rolled up an impressive record of 60,000 accident-free flight hours. New types of aircraft were added to the air station inventory: MACG-48 received UH-1N helicopters, and the VR Squadrons received C9B aircraft. The 1986 air show celebrated the 75th anniversary of naval aviation.

By 1987, the numbers had grown to 1,575 on active duty, 300 civilians, and 4,200 reservists. Even after the decision to close the base was final, work continued on buildings under construction—contracts had to be fulfilled. The air station sported new family housing and new bachelor enlisted quarters. Several facilities were virtually new when the base closed. A 1993 article that appeared in the *Glenview Announcements* confirmed the closing of NASG. It quoted the following figures: value of NASG: $280,000,000; acres of land: 1,288; active duty personnel: 1,642; reserve personnel: 2,954; civilian personnel: 369; aircraft assigned: 93.

Air shows and expos had been held on a regular basis after World War II and were always popular and well attended; the final event in 1993 was perhaps the most memorable, drawing 70,000 attendees. Hundreds of cars and onlookers parked along streets and roads adjacent to the base to watch a dazzling array of aircraft. *The Last Air Show*, a privately produced video, commemorated the event.

One unit, the Coast Guard, petitioned the Village of Glenview to remain on site. Since it was the only primary search and rescue unit to serve in southern Lake Michigan, the pleasure boating community and merchant ships depended heavily upon the Coast Guard. The request was denied; the unit was relocated to Traverse City, Michigan. During the years the unit was based at Glenview, crews flew more than 2,000 missions and saved 300 lives.

In August 1994, crews from the Marine Aerial Refueler Transport Squadron VMGR-234 flew the last KC-130 out of Glenview, bound for their new base at NAS Carswell in Fort Worth, Texas. Originally an attack squadron, VMA-234, it was activated during the Korean War. The squadron made history when, in December 1986, a KC-130T was the first marine aircraft to land on an ice runway at McMurdo Station, Antarctica.

The final weeks of base operations were a time for reflection. The country and the military had undergone vast economic and sociological changes. Founded at the beginning of the Great Depression, Curtiss Reynolds Wright had, against the odds, survived. The old airfield gave way to NASG, which played a crucial role during World War II. It provided a base of operations for those who served in succeeding wars and conflicts: the cold war, Korea, Vietnam, Desert Shield, and Desert Storm. The role of women in the military underwent unprecedented change at every level; by the time the base closed, there were women in the cockpits of base aircraft. While military fatalities are inevitable at facilities like NASG, a single civilian life was lost as a result of aircraft accident during the 58 years of flight operations.

NASG had a long tradition of public service. Over the years, base personnel flew humanitarian mercy missions, donated blood when medical crises occurred, and filled sandbags during flood conditions. Air station facilities were frequently made available to local schools, Boy and Girl Scout troops, the Glenview Council of the Navy League, veteran's groups, and civic organizations.

NASG was not simply a collection of buildings, runways, and aircraft; it was the officers and enlisted personnel (along with dedicated civil servants) who made the base a success. It has been a great pleasure to interview dozens of veterans who served there and to hear their stories. Their input was the basis of this book. Ideally, each and every story, anecdote, and picture could be included here—the stories of the heroes and rascals, the bravery and hijinks—are all were part of the mosaic that was NASG. Limitations of space dictate restrictions. Every individual and unit who served at the base was important. Some sea stories—such as the famous case of the misplaced P4Y Privateer—are perhaps best left for reminiscing at squadron reunions. Photograph captions include available names of those pictured; if identities are missing, that information has been lost to time.

Although this volume is not an attempt to present a comprehensive history of NASG, it is an effort to distill the essence of what the base was about and what it meant to those who were associated with it. Those who served at the base credit it with being a very important part of their lives. Glenview was not exclusively about military service, but about public service as well. Although each and every military facility has its own interesting history, NASG was one of a kind. The role it played in World War II was unique and monumental. Following that period, those stationed at the base continued to serve during the years of the cold war, Korean War, Vietnam, and Desert Storm. Career military personnel felt a special kinship with NASG; many retired to Glenview and surrounding communities. This is their story and a salute to all who served aboard NASG.

Resources consulted for this book include:
A Heritage of Wings by Richard C. Knott
Exhaust and *Glenview Views* (NASG newspapers)
Lake Michigan's Aircraft Carriers by Paul M. Somers
Naval Historical Center, Department of the Navy
Reflections by William Christensen
Some Early Birds: The Memoirs of a Naval Aviation Cadet, 1935-1945 by Joe Hill
The Final Salute, edited by Leslie Coker
The Last Airshow – Andries Communications and AVA Northwest, Inc.
The Life and Wit of a Navy Original: Admiral Dan Gallery by C. H. Gilliland and Robert Shenk
Top Guns of '43 – WTTW, Channel 11, Chicago
United States Naval Air Stations of World War II, Volume One by M. L. Shettle Jr.
U.S. Naval and Marine Corps Reserve Aviation, Volume One by Wayne H. Heiser
Whatever Happened to Curtiss Wright? by Robert W. Fausel

One

IN THE BEGINNING

Although flying machines proved themselves during World War I, there remained a great deal of skepticism. In the same way many Americans could not see much of a future for the motorcar several decades earlier, a majority of navy brass could not envision a time when airplanes would be an asset to the fleet. The Naval Militia Act of 1920 ended the historic volunteer naval militias and called for the formation of Navy and Marine Corps Reserves. Rear Adm. William A. Moffat became the first chief of the Bureau of Aeronautics in 1921; with his enthusiastic support, reserve aviation programs were organized.

In September 1923, the United States NRAB Great Lakes, Illinois, was established under the command of Lt. Richard E. Byrd, U.S. Navy (later Admiral Byrd, of arctic exploration fame). The training program at Great Lakes primarily used seaplanes and, later, a few land-based aircraft. The program was poorly funded and would not have been possible at all during the Great Depression without the support and cooperation of the civilian community. P. K. Wrigley, head of Chicago's vast Wrigley empire, and other wealthy civilians donated planes to the Great Lakes program. Those who served in the reserve were primarily men who lived in the Chicago area; most of the pilots flew for commercial airlines. Drill pay for reservists was unknown in that era, and pilots often had to reach into their own pockets to pay to fuel their training planes.

As heavier and faster planes were designed and built, the short runway at Great Lakes was unable to accommodate the next generation of naval aircraft. An exhaustive survey was undertaken to determine which locations in the Chicago area might be suitable for relocation of the NRAB. On November 1, 1939, word was received at Great Lakes that the Navy Department had approved a lease on Glenview's Curtiss Reynolds Wright Airport.

Written on the back of the picture is, "This aerial view shows the original location of (squadron) VN-16 RD-9 at Naval Training Center Great Lakes. The hangar can be seen left center in the photograph with concrete apron in front. The large structure is the aviation mechanics school. Remnants of old streets, sidewalks, and buildings of World War I vintage are in the visible surrounding area. Note the lone aircraft parked near the hangar." The photograph was originally owned by Charlie Swasas.

Pilots prepare for a training flight in these FBC-4 fighter/bombers. Note the sailor/mechanic (in the background) inspecting the gun mount. (Courtesy of Great Lakes Naval Museum.)

One aspect of training for Great Lakes aviation personnel was the parachute drill. (Courtesy of Great Lakes Naval Museum.)

The caption on this 1920 photograph indicates that the building on the right is the Machinist Mate School. The men of Company G are "troubleshooting on motors." (Courtesy of Great

This picture postcard scene from around the early 1930s shows the seaplane basin on the shore of Lake Michigan at Great Lakes Naval Training Center. (Courtesy of Great Lakes Naval Museum.)

Lakes Naval Museum.)

Sailors tow an N3N seaplane out of Lake Michigan onto the beach at Great Lakes. (Courtesy of Great Lakes Naval Museum.)

On this roster of Great Lakes personnel, dated 1934, photograph owner Carl Garisso wrote, "These people reflect the original personnel at Great Lakes who supported the squadrons which came out on weekends. We were called the Stationkeepers or Active Duty Personnel."

An inspection at Great Lakes sometime during the early 1930s is probably of a class of graduating seaman apprentice sailors. The officers wear swords, indicating it is an official occasion; the ranking officer is a captain.

Two

FROM CORNFIELD
TO AIRFIELD

Chicago's main airport, called Municipal (now Midway), originated in 1927. Coal was the primary fossil fuel in those years. The air was filled with black smoke, and pilots complained of poor visibility. The Curtiss Flying Service saw an opportunity; if they could build an airfield away from the "smoke belt," commercial and private pilots would flock to their facility. With relatively flat terrain and few power lines, Glen View (as it was then spelled) was considered ideal. Nine farms totaling 325 acres were purchased at a total cost of $607,500. Thus was launched the project envisioned to become Chicago's premier airport. Earle Reynolds, a prominent Chicago banker, arranged financing for the project. Part of the deal was that "Reynolds" would be included in the airfield's name.

In 1929, a boom time for the aeronautical industry, the Curtiss-Wright Corporation was formed by the merger of several top companies in the aircraft industry: CW's grand plan included a world-class commercial hangar, state-of-the-art field lighting, turf runways, a training school for pilots and mechanics, and a fly-in country club. The hangar, designed by prominent Chicago architect Andrew Rebori, was to be the largest in the world at that time. Dedication ceremonies took place on August 20, 1929. Nine days later—August 29, 1929—marked the beginning of the Great Depression and changed everything. Few potential pilots and mechanics could afford to attend the schools. The fly-in country club never got off the ground.

Commercial airlines added a bit of much-needed revenue in the late 1930s. On Sundays during the month of June 1938, United Air Lines offered a 45-minute sight-seeing flight. For a modest fee, individuals could board a Mainliner and fly "along the picturesque shores of Lake Michigan." United also operated scheduled Mainliner service from Glenview to New York, via Cleveland and Philadelphia.

Although the airfield managed to survive during the depths of the Great Depression, the navy's decision to relocate the Great Lakes aviation program to Glenview likely saved Curtiss Airfield.

Equipment from Glenview-based Roseman Tractor Company levels land for the Curtiss Reynolds Wright Airfield in spring 1929. The company's owner, Joseph Roseman, was a well-known golf course architect. He designed several courses in the Chicago area, including the 36-hole Pickwick Golf Club, which eventually became the NASG's Station Links. The M. R. Lovgren Construction Company of Highland Park, Illinois, was a general contractor that installed a complex drainage system for turf runways.

Curtiss Flying Service was one branch of the giant Curtiss Corporation, founded by Glenn Curtiss. Although Glenn Curtiss and Orville Wright (Wilbur Wright died in 1912) carried on legal battles over patent rights for years, their names would eventually be linked as several major corporations in the aviation industry merged to form the Curtiss Wright Corporation. Shortly following the merger, Glenn Curtiss died in 1930.

Maj. Rudolph W. Schroeder was the project manager for construction of Curtiss Airfield. Well over six feet tall, he was known as "Shorty." Major Schroeder was a World War I pilot and set several altitude records during the 1920s. A well-respected member of the aviation community, he was hired to oversee construction of Curtiss Reynolds Wright Airfield. He and his family lived in a home adjacent to the airport.

In the 1930s, the main hangar (known during the navy years as Hangar One) was a world-class facility. It featured many innovations, including a promenade providing passengers and visitors a panoramic view of the airfield. The interior design allowed passengers to view mechanics at work on the planes; the hope was that this would help alleviate any potential fears of flying. The airfield's infrastructure included an elaborate drainage system and state-of-the-art lighting. Turf that could withstand the rigors of taildragger planes was developed for grass runways. The partially visible road is Shermer Road, a main thoroughfare between west Glenview and Northbrook. After expansion of the base in World War II, the portion of Shermer Road running through airfield property was closed to the public. The 36-hole Pickwick Golf Club can be seen at the top of the photograph. (Courtesy of Eric Lundahl.)

This detailed map shows several points of interest, including the location of Chicago's Municipal Airport. Main roads were few and far between; there were no freeways and interstates in that era.

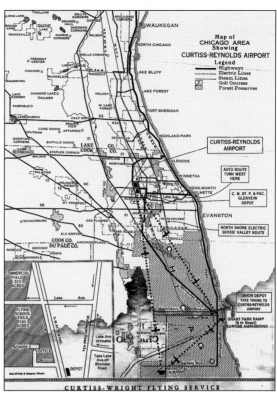

CURTISS-WRIGHT FLYING SERVICE

During the Great Depression, the $5 and $10 sightseeing tours advertised by Curtiss would have made a considerable dent in the average person's pocketbook.

About 10 months after its dedication in October 1929, Curtiss Airfield hosted the National Air Races, usually held in Cleveland. Military and civilian pilots competed on an equal basis. Fatalities among both pilots and spectators were not unusual.

Air show traffic was bumper-to-bumper, even in the 1930s. Cars head west on Glenview Road near the old Glenview State Bank. Curtiss Airport hosted the likes of Jimmy Doolittle, Charles and Anne Lindbergh, Eddie Rickenbacker, Roscoe Turner, and Amelia Earhart. The airfield was one of the destinations for the Ford Trimotor Tour; events such as these provided a much-needed financial boost for the airfield's owners.

The 1933 International Air Races were held at Curtiss. That same year, the Century of Progress World's Fair was underway in Chicago; many of those attending the fair also came to the air races. The Gordon Bennett Balloon Races were an added attraction. The Chicago, Milwaukee, St. Paul and Pacific Railroad added extra trains on the route from Chicago to Glenview. Seaplanes brought passengers from Chicago's lakefront north to Wilmette Harbor, where buses transported them to the airfield. Upwards of 60,000 people per day watched as a dazzling array of the most famous names in aviation appeared at the air races.

Wiley Post and his Lockheed Vega, *Winnie Mae*, were regular visitors to Curtiss Airfield. Will Rogers often accompanied Post and would sometimes sign autographs for those working on aircraft in the Curtiss hangar.

A ground crew refuels a Curtiss Fledgling at Curtiss Reynolds. Private aircraft, as well as commercial and military planes, used the facilities at Glenview. Some more frugal visiting pilots would hire a youngster and his little red wagon to fetch a five-gallon can of gasoline from a nearby service station.

The Chicago Girls Flying Club held air demonstrations and carnivals at Curtiss Reynolds Wright. Along with American Legion air shows, these provided much-needed revenue to keep the airfield afloat during the 1930s.

Three

THE NAVY COMES TO GLENVIEW

The recommendation for relocation of the NRAB to Glenview was approved by Adm. William D. Leahy (chief of naval operations); construction began early in January 1937. The navy leased the northern three-fifths of the hangar (soon to be known as Hangar One); modifications were made to add spaces for storerooms, as well as carpenter and engine overhaul shops. When the NRAB moved to Glenview in May, Lt. Comdr. G. A. T. Washburn assumed command of the new facility. Lt. James G. Sliney served as executive officer. Formal commissioning ceremonies for the NRAB at Glenview were set for August 28, 1937.

Among the requirements for cadet pilots were at least two years of college, between ages 20 and 28, and the ability to pass a rigid physical exam. Successful candidates were sworn in with the rate of seaman second class. NRAB Chicago was classified as an "E" (Elimination) base. Candidates received 10 hours of ground school and dual instruction. If a man could successfully solo after this training, he was eligible to go to Pensacola to pursue Navy Wings of Gold. Those who could not master the training were eliminated and might be offered alternative training.

In the early months there were no quarters at NRAB Chicago for station keepers, cadets, or reservists. Eventually space in Hangar One was set aside for aviation cadets. There were no mess facilities; meals were provided by "Ma" Bunge, who ran a boardinghouse about a mile east of the hangar. The food was said to be excellent.

During the late 1930s, training continued at a modest pace. As it became obvious that the United States would be involved in the war raging in Europe, the navy bought the entire property. According to the January 31, 1941, issue of the *Glenview View*, the Department of the Navy paid $530,000 for 319 acres. The same article indicated the property was worth more than $2 million.

After December 7, 1941, NRAB Chicago was never the same.

Lt. G. A. T. Washburn was the first commanding officer of NRAB Chicago. Appointed commanding officer of NRAB Great Lakes in 1936, he was recognized as a man with vision and the connections to get things done. These assets would prove valuable again at the beginning of World War II, when he was once again assigned to Glenview to oversee a massive expansion of the base.

Lieutenant Washburn's executive officer was Lt. James Sliney, whose many duties included recruiting. His home was located just west of the airfield, making it a convenient location to entertain visiting dignitaries. One of his duties was to take the show on the road to local colleges to recruit aviation cadets. His predecessor, Lt. Elmer L. Johansen, had been killed in a crash north of Curtiss Airfield, about one month before NRAB Chicago was commissioned.

The navy leased three-fifths of the Curtiss hangar. Extensive alterations were made to the structure to meet the needs of the new tenants.

It was customary to paint the name of an airfield on the roof of a hangar; this served as a navigation aid. After the navy moved in, the hangar roof was repainted to reflect the fact that the field was a combined civilian and navy operation.

COMMISSIONING

OF

U. S. NAVAL RESERVE

AVIATION BASE

CHICAGO

AUGUST 28, 1937

CURTISS AIRPORT

The program for NRAB Chicago's commissioning ceremonies listed squadrons from NRABs in Detroit, Kansas City, Minneapolis, Chicago, and St. Louis as participants in the accompanying air show. They demonstrated stunt, tactical, and formation flying and dive-bombing, as well as aerial combat. This was the first of a plethora of navy air shows to be seen in the skies over Glenview.

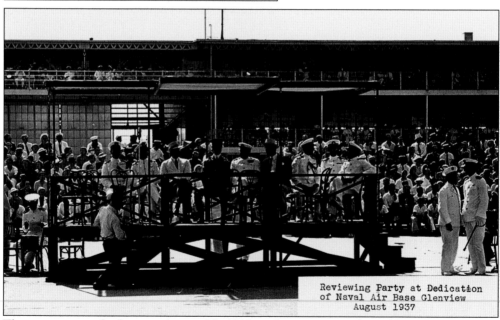

Reviewing Party at Dedication of Naval Air Base Glenview August 1937

The ceremony took place on August 28, 1937, and was open to the public—the crowd was estimated at 26,000. Dignitaries were seated on a covered platform in front of the Curtiss hangar. Among those attending were Rear Adm. Hayne Ellis, commandant of the 9th Naval District; army major general Frank Parker; and Ralph Church of the 10th Congressional District of Illinois. The price tag for the navy's new facility was $100,000.

This aerial perspective of the Curtiss/navy hangar and airfield reveals the method of training for bombing practice. Note the circle visible in the right lower portion of the picture. The circle was the target, and the bomb was a bag of flour that was dropped over the side of an open cockpit aircraft.

The first insignia used by NRAB Chicago and Squadron VS9R was designed by *Popular Aviation* magazine's art department. Featured is a graphic depiction of the Great Lakes superimposed on a bomb. Some 50 designs were submitted; final approval for adoption of this one came from the Navy Department in Washington, D.C.

Assistant Secretary of the Navy Charles Edison conducts an inspection of NRAB Chicago in 1938. A son of Thomas Edison, he later served as governor of New Jersey.

An impressive array of vehicles sits in front of the Curtiss hangar. Bill Christensen, one of the original station keepers at Glenview, recalls, "Depending on the dignitary to be impressed, all rolling equipment was polished and put out for base inspection."

In addition to the main Curtiss hangar, the navy had its own building at the north end of the airfield. This photograph shows officers and men of NRAB Chicago in front of the new purpose-built hangar. Bill Christensen, one of NRAB Chicago's "originals," remembers those who worked in the metal and engine shops used native intelligence and ingenuity to maintain aircraft when replacement parts and materials were scarce or nonexistent. Jack Witten, who served at the base in the late 1930s, recalls, "the work day was six days a week; the duty schedule was . . . two four-hour watches every other day. We guarded water works, radio stations [reference is to radio station WBBM's nearby transmission facility], power substations, and other facilities vital to the continued operation of the base."

The machine gun mounted in the rear of this fighter/bomber required that the gunner stand upright in the open cockpit to fire. The experience must have been unforgettable.

Naval Aviation Exhibit at International Air Show held at International Amiphitheatre, Chicago February 1938

Personnel from the base appeared at major events to publicize the navy and inspire young men to enlist. Here the Naval Aviation Exhibit appears at the International Air Show at the International Amphitheater in Chicago in February 1938.

Lt. (jg) Hyde and Mr. Ken Barton broadcas[t]
from plane during one of the series of No[tre]
Dame Squadron broadcasts.   August 1, 1[

As the need for pilots increased, so did recruiting programs. In the years prior to Pearl Harbor, regular radio broadcasts from NRAB Chicago were designed to pique the interest of potential aviation cadets. Special efforts were made to recruit college men—either graduates or those having attended for at least two years—to sign on for aviation training. Several squadrons of college alumni were formed, each taking the nickname of their respective schools. The Flying Irish (Notre Dame), Flying Illini (University of Illinois), Flying Wildcats (Northwestern), and many others transitioned from football field to airfield. From the beginning, a superior performance was the minimum requirement at Glenview. Station keeper Jack Witten recalls, "We taught them to fly, to navigate, to communicate, to recognize the enemy and find their way back to the carrier at night and in storms. We built their bodies in rigorous physical training . . . and taught them how to survive."

At the end of a two-week drill period in 1939, this squadron poses for a photograph. The plane is the base's J2F-4 Duck.

From left to right, in this photograph dated November 16, 1940, Lieutenant Commander Richards, Admiral Downes, Lieutenant Commander Gaines, and Commander McGinnis proudly display the Conway Trophy awarded to NRAB Chicago. The sterling silver trophy was a memorial to Lt. Edwin Conway, who died in a plane crash in 1933. He had been commanding officer at Floyd Bennett Field in New York City. The annual award went to the NRAB that received the highest final merit rating from the Naval Reserve Inspection Board.

Four

WORLD WAR II

The Japanese attack on Pearl Harbor in the early hours of December 7, 1941, galvanized the people of the United States in an unprecedented fashion. Even as the United States declared war on Japan, young men began lining up at local recruiting stations, ready to enlist in military service.

Plans were quickly set in motion to expand facilities at NRAB Chicago. In March 1942, Lt. Comdr. G. A. T. Washburn was ordered back to Glenview and assumed command of the base. At the time, there was a single barracks building, located north of Hangar One. With an appropriation of $12.5 million, Lieutenant Commander Washburn set to work to build a naval air training facility capable of turning out large numbers of pilots and crews. NRAB Chicago was officially designated a Naval Air Primary Training Command on October 1, 1942. Training not only continued but actually increased during the period of construction. Cadets received 100 hours of instruction (up from the previous level of 12 hours). On New Years Day, 1943, the Department of the Navy officially redesignated the base as a naval air station.

In the midst of the flurry of construction, another project was on the drawing board—the Carrier Qualification Training Unit (CQTU). It was the brainchild of Comdr. Richard Whitehead to convert two Great Lakes steamers into training carriers to qualify pilots to serve with the fleet. On August 22, 1942, the former luxury Great Lakes steamer *Seeandbee* was commissioned USS *Wolverine*. CQTU operations moved along quickly, and the first pilot successfully completed the required eight takeoffs and landings on September 12. Over the next several months the *Greater Buffalo* underwent refitting and on May 8, 1943, was commissioned the USS *Sable*.

May 8, 1945, marked the end of the war in Europe. War in the Pacific finally ended with an armistice with the Japanese three months later. Local residents, who were accustomed to the sounds of aircraft engines emanating from the planes flying overhead day and night, got used to newly quiet skies over Chicago's northern suburbs.

View of audience during the playing of the National Anthem at conclusion of Navy Day ceremonies at
Glenview, Ill. 27 Oct. 1941.

A few short weeks before the attack on Pearl Harbor, a celebration of Navy Day was held in Hangar One. Here on October 27, 1941, the audience stands during the playing of the national anthem at the ceremony's conclusion. This would be one of the last normal events held in the hangar for the duration of World War II.

It was clear in the early days of World War II that the facilities of NRAB Chicago would be inadequate to meet the rapidly escalating need for pilots and crews. Accordingly, the Department of the Navy allocated funds for a major expansion.

THE NAVAL AIR STATION IN 1942

An aerial view of NRAB Chicago from early 1942 shows few buildings, runways, and streets. Beginning in April of that year, an almost unimaginable transformation was about to occur. In the span of 121 working days, 1.3 million square yards of concrete were poured to construct landing mats and runways. The golf course, formerly 36 holes, was reduced to 18 in order to extend runways.

Civilian contractors were hired to build a self-contained community. Barracks, classrooms, mess halls, roads, administration, supply and commissary buildings, and hangars seemed to go up almost overnight. A recreation building was constructed to house a library, chapel, theater, game rooms, general store, and laundry, as well as tailor, barber, and shoe repair shops. A Drill Hall, complete with large training (swimming) pool, a hospital, and fire house were built. Finally, streets were paved, sidewalks were built, and streetlights and fireplugs were added. By the end of November 1942, the planned construction was essentially complete.

This 1850 farmhouse was the homestead of the Rugen family. Their patriarch, Herman Rugen, was one of Glenview's earliest settlers. He was the father of 13 children, one of whom owned the home at the beginning of the war. The navy, desperate for officer housing, took over the home and gave the Rugen family two weeks to vacate the property. When Rugen complained, the base commander reminded Rugen that "There is a war going on." Rugen replied that he was well aware, since four of his boys had just received draft notices that week. The Rugens had donated a small portion of their farm for construction of a small school to educate children from the west side of the village of Glenview. The navy notified Glenview's school board in May 1942 that the Rugen School would have to be vacated by June 1 of that year to make way for the base's south gate. Glenview's board of education pondered whether to move the building; in the end, they opted to demolish it.

The farmhouse, located near what became the south gate of the NRAB, was moved onto base grounds and remodeled to become the quarters of the commander. After closure of the air station in 1995, the house was demolished.

NRAB Chicago was transformed into a self-contained community, complete with new living quarters (barracks), hangars, shops, training buildings, recreation facilities, a hospital, a firehouse, roads, sidewalks, and fireplugs.

In this aerial view of the base, concrete aprons and a water tower are visible. Most of the buildings lining the streets were built between April and November 1942.

The swimming pool, located in the Drill Hall, provided both recreation and training. Many a recruit learned to swim here.

A rail spur from the main Chicago North Western Railroad provided a link for bringing in fuel and other supplies to the base.

A staggering amount of concrete was used to build the base's runways and circular landing mats. In all, 1.3 million square yards of concrete were poured in 121 working days. According to the navy, the job was one of the largest and fastest ever accomplished in the Midwest. (Courtesy of Eric Lundahl.)

The former NRAB Chicago was officially redesignated NASG on January 1, 1943. Now a center for both primary and carrier qualification training, the base covered more than 1,200 acres with greater than 350 airplanes in the inventory. Some 300 officers, 3,500 enlisted men, and 1,000 cadets were attached to the primary training command.

Comdr. G. A. T. Washburn turned over command of NASG on April 30, 1943, to Lt. Comdr. Truman Penny. His tireless efforts, ability, and planning made possible the transformation of NASG. Training not only continued but actually increased during the construction period: cadets received 100 hours of primary instruction (up from the previous level of 12 hours). The esteem in which he was held by his colleagues is demonstrated by this plaque presented to him at his departure. (Courtesy of Eric Lundahl.)

Commander Whitehead, Aviation Aide a
Lieut. Comdr. Gaines, Commanding Offi
Nov. 24, 1941.

As an aviation aide to staff Lieutenant Commander Gaines (right), commander of the 9th Naval District, Comdr. Richard F. Whitehead originated the idea of converting lake steamers to training carriers. The need for carrier pilots was particularly urgent in the Pacific, where planes and carriers were to play a major role in the war with the Japanese. It was a brilliant concept to establish a CQTU on the safe inland waters of Lake Michigan. Commander Whitehead's concept was adopted, and two side-wheel coal-fired lake excursion steamers, the SS *Seeandbee* and the SS *Greater Buffalo*, were identified as suitable to be refitted as training carriers. Shortly after the CQTU was operational, Commander Whitehead was reassigned to the Pacific Fleet, where he commanded the carrier *Shangri-La*. When the ship returned to port in Long Beach, California, at the end of the war, he arranged for a reunion of men who served aboard the *Sable* and *Wolverine*. (Courtesy of Eric Lundahl.)

The majestic Lake Michigan steamer *Seeandbee* was launched in 1913. A side-wheeler of magnificent proportions, its staterooms, dining rooms, and entertainment areas were beautifully appointed. The ship was a favorite for a generation of Midwest vacationers. (Courtesy of Lake County Discovery Museum.)

The SS *Greater Buffalo*, launched in 1923, was 10 years newer than the *Seeandbee*. It was one of the largest side-wheel passenger boats to ply the waters of the Great Lakes. In addition to passengers, it transported cars and express freight. During the Great Depression, steamers like this provided affordable lake cruises. (Courtesy of Lake County Discovery Museum.)

U.S.S. WOLVERINE
AUGUST 12, 1942

THE FIRST
ROSTER OF OFFICERS

About four months elapsed from the date of the *Seeandbee*'s purchase until it was commissioned as the USS *Wolverine* on August 13, 1942. The actual conversion from lake steamer to training carrier took less than two months. The first pilot successfully completed the qualification requirement of eight successful takeoffs and landings on September 12 of that year.

U.S.S. WOLVERINE

Training operations began aboard USS *Wolverine* almost immediately. The first aircraft was launched from the carrier on August 25, 1942. Once a pilot located the ship, he was amazed how small the deck appeared from his vantage point aloft. More than one man is reported to have thought, "you want me to land this plane on that postage stamp?" Although the prospect must have been daunting, more than 17,000 navy and marine pilots were successful.

U.S.S. SABLE AGAINST CHICAGO SKYLINE

The USS *Sable* was commissioned in May 1943. The USS *Wolverine* and USS *Sable*, along with their support vessels, formed the Cornbelt Fleet. Berthed at Chicago's Navy Pier, they sailed north on the waters of Lake Michigan each morning to receive planes from the base. Launch and recovery operations continued all day, usually seven days per week.

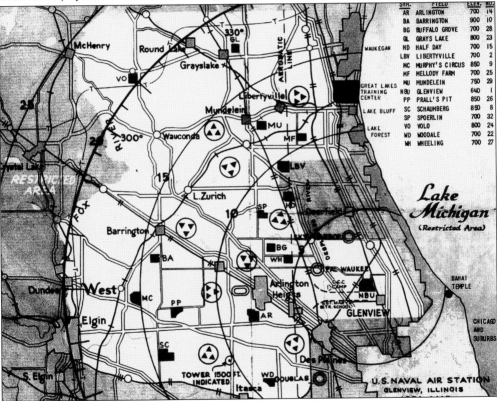

SYM.	FIELD	ELEV.	NO.
AR	ARLINGTON	700	14
BA	BARRINGTON	900	10
BG	BUFFALO GROVE	700	28
GL	GRAYS LAKE	800	23
HD	HALF DAY	700	15
LBV	LIBERTYVILLE	700	2
MC	MURPHY'S CIRCUS	850	9
MF	MELLODY FARM	700	25
MU	MUNDELEIN	750	29
NBU	GLENVIEW	640	1
PP	PRALL'S PIT	850	26
SC	SCHAUMBERG	850	8
SP	SPOERLIN	700	32
VO	VOLO	800	24
WD	WOODALE	700	22
WH	WHEELING	700	27

NASG had 15 outlying and bounce fields, which were built for it to use as satellites; some were alternate landing sites for planes in trouble. Others were configured to resemble carrier decks, complete with arresting wires. This allowed pilots to practice on dry land before attempting the more complicated process of landing aboard a carrier. Additionally, deck crews could learn their duties here before having to perform under more stringent conditions. It was designed to be so effective that men could make a nearly seamless transition from training to sea duty.

The oldest building aboard NASG, Hangar One's official name was Building One. Although only a dozen years old when the United States entered World War II, extensive changes were made to both the interior and exterior. The hangar served as the nerve center for both primary and carrier training units.

Cadets wait their turn to fly in this image. Rosters that listed names and flight times were posted on the chalkboard seen in the background. By April 1943, there were 350 planes, 300 officers, 1,000 cadets, and 3,500 enlisted personnel attached to the primary training program.

"Yellow Peril" was the nickname for bi-wing primary training planes, reflecting the bright yellow paint scheme. The biplanes were either N3Ns or Stearmans. N3Ns, very similar to Stearmans, were manufactured in the navy's own aircraft factory located in Philadelphia.

After taking off from NASG, pilots would fly east toward Lake Michigan. The Baha'i House of Worship in Wilmette, known as Point Oboe, was the first destination. Here squadrons would rendezvous and fly out over the lake to locate the carrier to which they had been assigned. Having made radio contact with the ship, each would take his turn to attempt the required eight successful takeoffs and landings on the carrier's deck. (Courtesy of the Baha'i House of Worship.)

Deck crews, as well as pilots, received vital training aboard the carriers. Here a "yellow shirt" (a term referring to a deck crewman who guides a pilot to launch) uses hand signals to communicate with the pilot of this TBM torpedo bomber. Other aircraft used by the CQTU included SNJs (Harvards), F4Fs (Wildcats), SB2Cs (Helldivers), SBDs (Dauntless), and SNCs (Falcons).

Not all landings were successful. Engine failure, fire, or pilot error could result in a plane going into the water. Pickup boats were always nearby, and the pilot's chances of survival were excellent.

The job of a deck crewman was hazardous. Broken arresting wire cables were constant concerns. An out-of-control plane could careen across the deck, striking anyone in its path. Fires could also break out. A moment's lapse in concentration could result in a man walking into whirling propeller blades. Accidents could, and did, change lives in an instant. (Courtesy of Eric Lundahl.)

Because Hangar One was the center of operations, it was a beehive of activity. Pilots and crews gathered to celebrate successful qualifications or to commiserate when things did not go so well.

Ens. Charles S. Downey was the youngest naval aviator in World War II. Earning his wings at age 18 in 1943, he completed CQTU at Glenview, then was assigned to the USS *Ticonderoga*. He was decorated with the Distinguished Flying Cross for heroic service in the Pacific. Downey remained in the postwar Naval Reserve. When he was promoted to captain at age 39, he was the youngest man in the history of the Naval Reserves to attain that rank. (Courtesy of Capt. Charles Downey.)

Movie actor Robert Taylor served as a flight instructor at Glenview. Taylor, a lieutenant, also starred in training films for aviation cadets. Here Taylor (far left) poses with navy pilot Richard Nelson (far right), who did the flying in the film. Wilding Studios was in charge of production.

The most famous pilot to qualify aboard the *Sable* was George H. W. Bush. A few days older than Downey, he was the second youngest navy pilot of World War II. Lieutenant Bush is shown here flanked by TBM Avenger (torpedo bomber) crewmen. Assigned to the USS *San Jacinto*, his plane was shot down in 1944, and he was rescued by a submarine. Among his decorations was the Distinguished Flying Cross. He later became the 41st president of the United States.

GNAS 1st Insignia - V for Victory.

NAVAL AIR STATION GLENVIEW, ILL.

Several designs were submitted for a new patch for NASG. Named the winner in April 1943, "Fighting Cock" was designed by Ptr2c E. R. Howard of Indianapolis. He had operated a sign studio before the war. The prize was a $25 war bond.

In the 1930s and very early 1940s, firefighting equipment was simple. Although the base owned a fire truck, the modus operandi often consisted of a motorcycle with large tanks of carbon dioxide mounted in an attached sidecar.

U.S. senator Ralph Church (left), who represented the 10th and 13th districts of Illinois during his political career, was a staunch supporter of NASG. The naval officer pictured is Lt. Richard Schram.

It was the responsibility of plane captains to ensure that aircraft were in good condition. They would often start a plane and taxi it into position before turning it over to its pilot.

Admiral Carpenter and his party inspect the base machine shop.

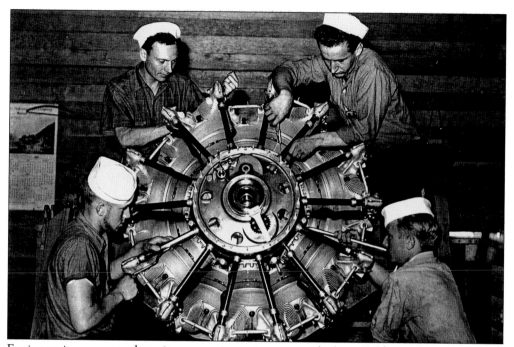

Engine maintenance and repair was a major assignment at the base. Publicity photographs such as this were used to show the folks at home how their boys were helping to win the war.

Women volunteered for military service during World War II in unprecedented numbers and proved themselves capable of handling jobs that would have been unthinkable assignments before the war. The first three WAVES came aboard NASG in May 1943. Their assignments varied from yeoman (clerk) to ground school instructor to mechanic. For some of their male counterparts, the idea of women doing "man's work" took some getting used to. By December 1944, there were 322 enlisted women and 12 officers attached to the base.

The NASG's band had a large role in keeping spirits up during the war years. They played frequent concerts, and smaller groups formed dance bands. Contingents of young women from nearby suburbs attended base dances, which were great morale boosters. In addition to the band, there was an orchestra composed of members who, before the war, had played with such illustrious groups as the Chicago Symphony and the Philadelphia Orchestra. Additional social functions were organized by local families, who invited base personnel to Sunday dinner and collected books, furniture, and table games for the air station's recreation areas.

EXHAUST

NAVAL AIR STATION-GLENVIEW, ILL.

VOL. IV   NO. 5   I MARCH 1944

Security was of paramount concern at the air station; sentry K-9 units were formed to help with patrols. In 1943, NASG received some of the first Dogs for Defense canines from the War Dog Reception and Training Center in Virginia. No information is available as to how the dogs were outfitted for the cold Chicago winters. Their handlers, however, had the opportunity to buy long underwear made of wool and cotton; the base exchange advertised tops and bottoms for sale at $1.30 each.

An emergency in the air created a crisis for pilots and crews. Even in non-combat circumstances, engines could fail or catch on fire. Riggers who worked in the parachute loft took their work very seriously, knowing that lives depended upon their expertise.

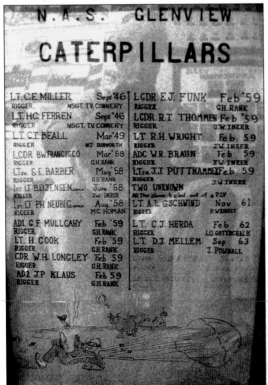

N.A.S. GLENVIEW

CATERPILLARS

LT. C.E. MILLER	Sept '46	LCDR. E.J. FUNK	Feb '59
RIGGER, MSGT. T.V. CONNERY		RIGGER G.H. RANK	
LT. H.C. FERREN	Sept '46	LCDR. R.T. THOMMEN	Feb '59
RIGGER MSGT. T.V. CONNERY		RIGGER J.W. INZER	
LT. C.T. BEALL	Mar '49	LT. R.H. WRIGHT	Feb 59
RIGGER W.T. DUNWORTH		RIGGER J.W. INZER	
LCDR. B.W. FRANCISCO	Mar '68	ADC. W.R. BRAUN	Feb 59
RIGGER G.H. RANK		RIGGER J.W. INZER	
LT jg. S.E. BARBER	May 58	LT jg. J.J. PUTTKAMMER	Feb 59
RIGGER G.H. RANK		RIGGER J.W. INZER	
1st LT B.D. JENSEN	June '58	TWO UNKNOWN	
RIGGER J.W. INZER		All the above bailed out of a P2V	
1st LT PH. NEUBIG	Aug '58	LT. A.L. GSCHWIND	Nov 61
RIGGER MC HOMAN		RIGGER P. WENNET	
AD1 G.F. MULLCAHY	Feb '59	LT. C.J. HERDA	Feb 62
RIGGER G.H. RANK		RIGGER LG GOTTSCHALK	
LT. H. COOK	Feb '59	LT. D.J. MELLEM	Sep 63
RIGGER G.H. RANK		RIGGER I. POWNALL	
CDR. W.H. LONGLEY	Feb 59		
RIGGER G.H. RANK			
AD2 J.P. KLAUS	Feb 59		
RIGGER G.H. RANK			

Those who successfully bailed out of an aircraft became members of the Caterpillar Club. It was customary for the individual to present a bottle of whisky to the rigger who had packed his parachute.

The Drill Hall was in constant use for ceremonies as well as training. The spacious building was also ideal for dances and other social gatherings. (Courtesy of Eric Lundahl.)

Rear Adm. Edward C. Ewen (front row, center) appears with his staff in front of the Naval Air Reserve Training (CNAResTra) Headquarters Building at NASG. Admiral Ewen assumed command in December 1945, although the CNAResTra designation did not become official until July 1946. One of the main streets at the air station was named in his honor.

WAVES Third Anniversary
86,000 Strong!

In August 1943, NASG WAVES posed in front of the administration building to commemorate the third anniversary of the organization of the corps. Letters of congratulations were received from Secretary of the Navy James Forrestal and Fleet Adm. Ernest J. King. NASG commanding officer Capt. J. N. Carson extended the traditional navy "well done" to the women.

Celebrities did their bit for the war effort in a variety of ways. Here movie actress Dorothy Lamour poses for a publicity shot at an all-star football game in Chicago. It was not unusual for players to be sworn into the service following such games. Such events served as inspiration for others to join the war effort.

Although the little Yellow Perils were most visible in the skies over Glenview during the war years, flights of another nature were also part of operations at the base. In the early months of 1944, the Naval Air Transport Service was activated at NASG. Two flights were flown daily from the base—one eastbound and the other westbound—with a cargo of aircraft parts and lightweight precision tools manufactured in the Great Lakes industrial region. The essential supplies were destined for factories from coast to coast.

The pace at Hangar One slowed considerably by mid-1945. Primary training continued, and yellow biplanes still flew over the tower, but in far fewer numbers than in previous months. In all, some 10,000 primary pilots and more than 17,000 carrier pilots were trained.

Allied success in Europe and the Pacific heralded the need for fewer pilots in the final months of the war. Accordingly, the CQTU program was phased out at NASG. No longer needed for operations, the *Sable* and *Wolverine* were decommissioned on November 7, 1945. Operations statistics kept during the war revealed an impressive 786,982 daytime flying hours, 27,465 nighttime hours, and approximately a quarter million takeoffs and landings. Ens. C. E. Gilette of Dillberg, Pennsylvania, made the last landing on September 15, 1945. Here, berthed at Navy Pier, the carriers await their fate. The *Wolverine* was scrapped in November 1947, and the *Sable* followed several months later in 1948. (Courtesy of Capt. William McConnell.)

Five

POSTWAR AND KOREA

Less than a month after the armistice with Japan was signed, attention turned to planning for naval aviation in the postwar era. On September 5 and 6, 1945, five rear admirals, all pilots, met at NASG to formulate recommendations for basic training and number of pilots necessary for peacetime operations. Their work led to a plan for maintaining a sizeable number of experienced pilots, mechanics, and aircrews. Planes and support equipment were turned over to reserve units. Because of its central geographic location, NASG was designated headquarters for navy and marine reserve aviation training commands in 1946. Combat-experienced pilots and crews formed the nucleus of these units; they trained on weekends and came to be known as "Weekend Warriors." At Glenview, navy and marine units used the same planes; they alternated drilling weekends so that one weekend per month, NASG functionally became a marine reserve air station. By 1950, there were 16 squadrons and 2,000 reservists based at Glenview. An additional 2,500 volunteer reservists drilled there as well.

The Korean War began on June 25, 1950. Thirty thousand members of the Naval Air Reserve were called upon to augment active navy personnel as mothballed aircraft carriers and support vessels were quickly returned to service. Since NASG was headquarters for both naval and marine air reserve commands, the pace at the base once again accelerated. NASG reserve squadrons deployed to Korea included NAR-702 and VF-721, VF-725, VA-728, and VF-727; all these were carrier-based. The first units to arrive for duty in South Korea were attached to Carrier Group 15, which included the USS *Boxer*. Squadrons arriving later were assigned to Carrier Group 77; this group included the USS *Antietam*.

War had been waging for several months when marine fighter squadron VMF-121 was activated in March 1951. They joined Marine Ground Unit Radar Intercept Squadron 22, which had been called up the previous August. The fighter squadron was land based, flying F4U Corsairs. Other marine units activated were Marine Air Control Group 48 (MACG-48) and Marine Air Squadron 22.

With cessation of hostilities in July 1953, navy and marine reservists returned to Glenview and resumed weekend training drills.

At the ceremony pictured here, Comdr. H. E. Benefield relieved Lt. R. A. Schram as the naval air station's A and R (assembly and repair) officer. Lieutenant Schram, who had served at the base for three and a half years, received seven commendations during that time.

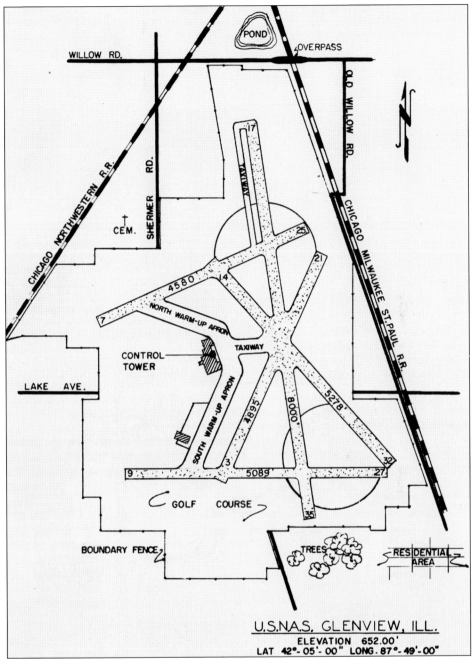

POND

OVERPASS

WILLOW RD.

OLD WILLOW RD.

CHICAGO NORTHWESTERN R.R.

SHERMER RD.

CEM.

17

TAXIWAY

CHICAGO MILWAUKEE ST. PAUL R.R.

25

21

4580

14

7

NORTH WARM-UP APRON

TAXIWAY

CONTROL TOWER

LAKE AVE.

SOUTH WARM-UP APRON

4895

8000

5278

3

5089

9

27

32

35

GOLF COURSE

BOUNDARY FENCE

TREES

RESIDENTIAL AREA

U.S. N.A.S. GLENVIEW, ILL.
ELEVATION 652.00'
LAT 42°- 05'- 00" LONG. 87°- 49'- 00"

Maps of United States air stations were a routine part of each pilot's flight gear. This 1950s map of NASG contains information about runways and taxiways, as well as roads and rail lines adjacent to the field. The circular mats were for the use of small training planes, which required short distances for takeoffs and landings. Short runways were used by small planes, and the longer runway for larger, heavier ones. Warm-up aprons are visible on the left, to the north and south; here pilots warmed up engines and awaited clearance to move onto the runway for takeoff. Runway 9/27, the emergency runway, was equipped with a tail hook and resist apparatus to slow a disabled plane.

On July 1, 1946, NASG discontinued primary flight training and officially became headquarters for the naval air reserve training command. Some 11 months following the victory over Japan, the navy's reserve training program was in place. Shortly thereafter, Glenview also was designated as headquarters for marine air reserve training.

The marine air reserve training command headquarters was located across the street from the navy's large building. Housed in a former barracks, it was a far less impressive structure. The marines were said to be happy with it, since they were accustomed to roughing it. (Courtesy of Col. A. C. Realie.)

With Glenview now headquarters for navy and marine air reserves, a navy admiral and marine general were aboard as commanders of their respective units. A new insignia was designed to incorporate the flags of the marine and navy commanders. The base was under the command of a navy captain.

When the Admiral's resided at GNAS.

This view of the main (west) gate of the base was taken around 1950. In the center of the sign is the insignia that identifies NASG as headquarters for navy and marine air reserve training commands. Other units were also based at the air station, including the navy's Combat Information Center. Established in 1947, it was the largest radar aircraft interception school anywhere at the time and drew students from foreign countries in addition to those from all branches of the United States military services.

VMF-121 pilots report for duty during a drill weekend at NASG. The chalkboard in the background is similar to those used during World War II to list duty assignments. Squadron commander Capt. A. Charles Realie (second from left) briefs pilots on the day's operations. The pilots are, from left to right, Raymond Teplis, Herschel Watson Jr., and Arthur W. Kamp. This rigorous training paid off when the squadron was activated for duty in Korea.

Members of Marine Squadron VMF-121's line crew take time out for a family photograph. The men were drilling reservists at Glenview prior to the Korean War and then were deployed overseas when their squadron was activated. (Courtesy of Col. A. C. Realie.)

Marine pilots from NASG prepare for deployment to Korea on March 1, 1951. (Courtesy of Col. A. C. Realie.)

World War II pilot John J. Gauss, a Glenview reservist, was recalled to active duty and served with the first Marine Air Wing in Korea. (Courtesy of Col. J. J. Gauss.)

On October 11, 1951, the target was an airfield at Kyonhyang, Korea. Hit by antiaircraft fire, Captain Gauss's plane went down, landing in the Yalu River. This photograph was taken by a plane that was flying cover for the downed aviator. (Courtesy of Col. J. J. Gauss.)

Captain Gauss wades toward the shore, while carrying his parachute on one shoulder. It is interesting that he would have been concerned about his parachute, given the circumstances. Had it been lost, however, he would have been required to pay for it. He was rescued by a Grumman SA-16 Albatross. (Courtesy of Col. J. J. Gauss.)

Pilots and aircrews serving in Korea were issued a safe conduct pass like this. Its message, which was written in several languages, stated that anyone safely delivering a pilot or crewman to American lines would receive a reward. (Courtesy of Col. A. C. Realie.)

Pictured here underway off Korea, the USS *Antietam* (CV-36) was manned by reserve squadrons from Glenview, Denver, and New York City. In a typical month during the war, 8,000 combat sorties were flown by American carrier pilots—6,000 of them by reservists. There were lighter moments, however. The USS *Antietam*'s combat action reports for November 26–December 31, 1951, include a paragraph detailing a dilemma facing the carrier's post office. The volume of outgoing packages was described as having "swelled to enormous proportions," which was attributable to the availability of "cheap and novel goods in Japan."

Pilots from Squadron VA-728 are pictured on the deck of the USS *Antietam*. They are, from left to right, (first row) Lt. Glen Geho, Ens. William Bird, Lieutenant (j.g.) Egeland, Lt. (j.g.) Joe Voda, Ens. Marvin Braddock, and Ens. Robert Thomas; (second row) Lt. Joe Neri, Lt. Richard Thommen, Lt. (j.g.) Robert Parsons, and Ens. John Higgins; (third row) Lt. Carl Dorfler, Lt. James Walley, Lt. (j.g.) John Nicholas, Ens. Ben Hemphill, Ens. Howard Hochn, Lt. Charles Noth, Lt. Francis Dwyer, Lt. (j.g.) John Sbermolis, Lieutenant Arthur, Lt. Gavin Wier, Lt. Comdr. Wilbert Hackbarth, and Lt. Daniel Price.

VA-728 deck crewmen secure aircraft on the deck of the USS *Antietam*.

The Korean War ended in 1953. Many of those who served returned to civilian life and remained in the reserves. These marine pilots have returned to NASG following a training flight. (Courtesy of Col. A. C. Realie.)

Six

THE COLD WAR YEARS

Almost as soon as World War II ended, the cold war began. Historians cite the beginning of the cold war as 1947, when tensions between Russia and the United States reached major proportions. The two superpowers clashed over plans for rehabilitation of Germany and led to events such as the Berlin Blockade and Berlin Airlift. It was a very dangerous time in history, since nuclear war was a constant concern. It was clear that a ready reserve was necessary for rapid response, should the situation deteriorate.

The Berlin crisis resulted in activation of the VS-721 squadron. This squadron was deployed to Widbey Island, Washington, for one year. Both VS and VP squadrons were mobilized to fly ASW (antisubmarine warfare) missions in response to construction of the Berlin Wall in 1961. In January 1963, three officers of Air Wing Staff 72 served a two-week cruise (training exercise) during the Cuban Missile Crisis.

In late 1969, there were 11 squadrons based at Glenview—seven navy and four marine. When the Naval Reserve program was reorganized in 1970, most units were decommissioned. As a result, patrol squadrons VP-60 and VP-90 were officially formed on November 1, 1970. VP-60 had a nucleus of full-time, active duty personnel supplemented with reserve personnel. VP-90's personnel were both active duty and Select Air Reservists. At the same time, Fleet Tactical Support Squadron 51 Detachment, Glenview (VR-51) was commissioned as a Naval Air Reserve Force Squadron.

Major reorganizations occurred in 1973, when Naval Air Reserve Headquarters was moved to New Orleans and renamed Naval Air Reserve Force. Remaining at Glenview were Patrol Squadrons VP-60 and VP-90. Personnel included 450 men comprising 15 flight crews; their mission was antisubmarine surveillance. Fleet Logistics Support Squadron VR-51's mission was to support the Sixth Fleet, Rota, Spain, and Seventh Fleet, Guam.

By the late 1980s, the government began to close military bases all over the country. Many began to speculate about the future of NASG.

Hangar One continued to be the center of operations during the 1950s. Its appearance was virtually unchanged from the World War II years.

After NASG became headquarters for both Navy and Marine Air Reserves, this insignia was painted on the tail of planes based there. The images within the shield reflect air and ground personnel, with Hangar One and two aircraft visible in the arc below and Navy Wings of Gold superimposed at the bottom. A navy orientation publication for new NASG personnel stated, "Here operations are directed for training thousands of 'Weekend Warriors' through 12 Naval Air Stations, 6 Naval Air Reserve Training Units, and 5 Naval Air Reserve Divisions located near principal cities throughout the nation."

GNAS Base Insignia used on Aircraft.

Adm. Daniel Gallery, seen here inspecting personnel, was assigned as commander of Naval Air Reserve at Glenview in 1952. Having been skipper of the task force that captured the German submarine U-505, he was happy to have the vessel exhibited at Chicago's Museum of Science and Industry (MSI). Glenview's proximity to Chicago allowed him to promote the museum's acquisition of the U-boat. His efforts were successful, and U-505 is now on permanent display at MSI.

Comdr. Charles Downey (right) is shown here at the 1960 change of command ceremony for VA-722 with Comdr. Bobbie R. Allen. After returning to Glenview from duty in the Korean War, the squadron transitioned to Grumman S2F planes. The mission changed to antisubmarine warfare and was redesignated VS-721. The squadron was put to the test in 1959, during Operation Skynet. The future of the Naval Air Reserves depended upon the performance of the Weekend Warriors. To their credit, they came through with flying colors, turning in a better performance than many of the active duty units.

This two-seat, open cockpit Meyers biplane was owned by reservists Charles Downey, Mel Schmidt, Joe Duffy, Fran Moyer, and airline pilot "Smokey" Calahan. It was often flown at air shows to promote aviation and navy recruiting.

In addition to regular worship services, weddings were a frequent occurrence in the chapel. Here PH2 (photographer's mate second class) Eric Lundahl and Ramona Hocking are pictured with their wedding party in 1956. (Courtesy of Eric Lundahl.)

Pictured in front of Hangar One in early the 1950s, Frank Tallman (right) flew to NASG on a promotional tour for an oil company. Tallman was a navy pilot during World War II, having completed primary flight training at NASG. After the war, he and Paul Mantz formed the Tallmantz partnership. They owned a vast array of vintage aircraft and performed stunt flying for Hollywood films. Their credits included *It's a Mad Mad Mad Mad World* and *Flight of the Phoenix*. Both men ultimately died in plane crashes.

These F6Fs, parked in one of the bays of Hangar One, were carrier-based. Mechanics often worked long hours and late into the night to make necessary repairs so planes were ready to fly the following day.

The Drill Hall was the scene of many auspicious occasions. Here it is decked out for the annual inspection.

Skitch Henderson gives a concert at the base in the 1950s. Entertainers Danny Thomas, Jack Benny, and Fran Allison (later of Kukla, Fran and Ollie fame) had performed at the air station during World War II. In the 1960s, Ann-Margret (Olson) gave concerts at NASG before she became a movie star.

Distinguished visitors were an integral part of the action at NASG. Pres. Dwight D. Eisenhower, along with Mrs. Eisenhower, son John, and grandchildren, disembarks from the presidential plane that landed at NASG.

Jets began to appear at Glenview during the 1950s. Here an A-4 Skyhawk is parked on the apron in front of Hangar One. (Courtesy of Eric Lundahl.)

On a training flight over the city of Chicago, these F9F Cougars are an impressive sight.

The Douglas C-118 was the perennial workhorse of Glenview's Transport Squadrons. Most often used during the 1960s to transport reservists for weekend duty, they could airlift personnel to mobilization bases and augment operating forces in case of national crisis.

Helicopter squadrons were added to the aircraft inventory during the 1960s. In addition to general utility flights, their mission was placement of sonar buoys for tracking submarines.

Jacqueline Jacquet was one of about 120 navy flight nurses during World War II. She was stationed first at NASG and then was transferred to serve with the Naval Air Transport Service, evacuating wounded troops from islands in the South Pacific.

Lieutenant Jacquet was recalled for duty at Glenview during the Korean War. While serving at NASG, she met Marine Lt. Martin Melvin; they were married at El Toro Marine Air Station California, in 1950. Shortly thereafter, he was deployed to Korea with his Glenview-based squadron VMF-721. He returned safely, and the couple raised a large family, many of whom served in the military. Their daughter, Comdr. Maureen Christopher, also a navy nurse, was stationed at Glenview in 1981.

Hangar One housed numerous departments and offices. Log books were maintained in the crews' ready room.

Adjacent to the hangar bays were spaces for the storage of tools and maintenance equipment. Here a sailor checks aircraft parts.

Owned by Chief Petty Officer Tony Ruggio (right), this 1922 Buick carried scores of officers and petty officers into retirement at NASG.

A retiree receives the salutes of shipmates as he departs from his retirement ceremony.

It was not all work and no play at the base, as shown in this photograph of the Buick alongside a 9F9 Cougar.

Airships like these blimps often appeared at events to promote the navy and naval aviation.

The Flying Professor

Captain Dick Schram
United States Naval Reserve

Highly Cub-Sonic Comedy
Aerobatics with a Daredevil Flair

After World War II, Capt. Richard Schram performed at air shows throughout the country. Known as the Flying Professor, he piloted a yellow Piper Cub and performed a comedy routine, which became a favorite among air show crowds.

The Officers and Men of the Naval Air Station, Glenview proudly dedicate this publication to the Weekend Warriors now serving in the Korean area.

We here, salute you, and wish you Good Luck and Godspeed.

Dick Schram lost his life while performing at an air show in Reading on June 4, 1969. The NASG chapel was named the Captain Richard A. Schram Memorial Chapel in his honor. The chapel was built in 1944; local lore indicated that German prisoners of war, incarcerated in a camp nearby, provided some of the construction labor.

Chap. Bashford S. Power and Capt. Drexel E. Poynter participated in the dedication of Schram Memorial Chapel on January 26, 1970. One of the beautiful stained-glass windows in the chapel was dedicated to the memory of Mrs. Poynter.

A banner depicting the last base insignia is displayed in the chapel's sanctuary. In the center is a three-sided altar, built to revolve and reveal worship centers for Protestant, Roman Catholic, and Jewish services. The mechanism used to accomplish this was the type used to rotate a five-inch gun on shipboard.

A tenant service at NASG was the United States Coast Guard Station, which became operational on March 1, 1969. The facility was officially commissioned on March 15, 1969. Among the dignitaries present were Adm. W. J. Smith, commandant United States Coast Guard, and Rear Adm. W. F. Rea, commander of the 9th Coast Guard District.

The United States Coast Guard facility was located to the north on the air station property. A new hangar provided space for helicopter storage as well as room for aircraft maintenance and administrative offices.

Marine color guards from the base often led parades there, in the village of Glenview, and throughout the Chicago Metropolitan area.

Marine reservists from NASG were deeply involved in the 1964 Toys for Tots collection. Glenview resident Sgt. Lloyd Kuehn (front row, second from left) holds a toy guitar destined to make the holidays brighter for a Chicago-area youngster.

In September 1973, NASG's circle insignia was replaced by this triangular one. It was designed by AC2 James Lipski of NAS Operations. It was the final insignia and was identified with the base up to the time of closure in 1995.

Snow—sometimes lots of it—can make Midwest winters a challenge. The blizzard of 1979 left NASG runways buried under 35 inches of the white stuff. Front-end loaders helped clear it away.

Old-timers would remember the days before there was much mechanized snow removal at NASG. Modern equipment was a big improvement over the days when base personnel had to shovel runways, aprons, and streets by hand.

The Operations Office, located in Hangar One, was a familiar sight to anyone flying into or out of NASG.

Hangar One was also home to the Meteorology Office, which was considered one of the best in the fleet. Here pilots and air crews could check weather forecasts and related data to help them know what to expect en route.

Naval Reserve Aviation's own precision demonstration team, the Air Barons, was based at Glenview. They flew from 1969 through 1971, when federal funding for the program was discontinued.

The Air Barons' appearance at air shows was designed not only to entertain but to inspire young aviators.

In the early 1970s, these women were employees on the staff of the chief of Naval Air Reserve at Glenview. Two of them, Mary Fenoglio (front row, second from left) and June McGee (back row, far right), were veterans of World War II. Both served in the Coast Guard as SPARS (acronym for that service's motto Semper Partus—Always Ready).

The base offered many recreational opportunities. Bowling leagues provided friendly competition (as well as some exercise) during off-duty hours.

The marines finally got their own hangar in October 1977. The facility provided space for aircraft maintenance and some storage. A highlight in its history was the hangar dance held there, in conjunction with the final air show.

NASG air expositions were a big hit with the public. The popular annual events were hosted by NASG for nearly 45 years. Chestnut Street became a dead end at the east gate of the base, allowing families with children a safe spot and a good view of base runways.

At air expositions, visitors had an opportunity to get a close look at military planes that were seldom seen by the public. They also allowed the crowd to tour hangars, talk with personnel, and walk on the runways that were normally off limits. The shows drew huge crowds, not only from the Chicago metropolitan area but also from surrounding states. (Courtesy of Eric Lundahl.)

Demonstrations of the air and sea rescue equipment fascinated the public at base air show and open house events. In addition to performing the usual water-related rescues, the Coast Guard's unsung heroes plucked trapped construction workers from such unlikely places as smokestacks and water towers.

Demonstrations of firefighting equipment often took place at the expositions. Everyone—especially kids—loved to watch the big trucks in action.

The public was able to get a close look at a variety of military planes. Pilots and crews were on hand to answer questions and, sometimes, even sign autographs.

The distinctive British Vulcan bomber was a regular visitor to air station expositions. Tragedy struck in 1978, when it crashed on a routine training flight. Observers on the ground reported that the huge plane appeared to veer away from the populated areas surrounding the base and head toward open land north of the air station, where it went down. All four Royal Air Force crewmen aboard were killed. Investigation indicated that engine trouble led to the disaster.

The Blue Angels were based at Glenview when they were in the area to perform at Chicago's Air and Water Show and other events. From time to time, Glenview also hosted the Royal Canadian Air Force Snowbirds and demonstration teams from England, France, and Italy.

Thunderbirds visit Glenview

The Air Force Thunderbirds were often based at NASG when in the area for performances.

Pilots of *Patrouille de France*, a precision demonstration team of the French Air Force, return from a performance.

Fleet Logistics Support Squadron VR-51 prepares to depart from NASG for a cruise (training exercise) to Rota, Spain, in 1973. (Courtesy of Irving Danneil.)

One of the many programs that the navy sponsored for the public was Sea Cadets. This inspection of a Sea Cadet unit took place on May 4, 1978.

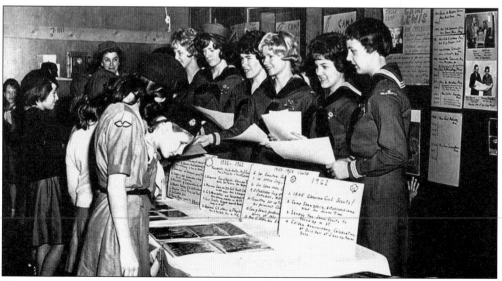

The navy provided sponsored events for girls as well. Exhibits highlighting Glenview-area Girl Scout activities are on display here, at an NASG open house.

Pictures on this and the opposite page are taken from the Chicagoland Plymouth Dealers' Model Airplane Contest, held annually at the Naval Air Station, Glenview. Each year, thousands of model airplane builders compete in the two-day meet. Thus, the Naval Air Station offers the youth of the Midwest an opportunity to develop their interest in aviation.

For many years, model airplane meets were an annual event hosted by NASG. This undated photograph, probably from the 1950s, was taken at the west (main) gate of the base. Signs indicate that the meet was cosponsored by the Chicagoland Plymouth Dealers.

The meets drew participants from all over the Midwest. They provided an opportunity for kids to see the air station in action, as well as further their interest in aviation.

During the Vietnam War years, MAC (aeromedical evacuation) flights arrived in Glenview on a regular basis. This Air Force C-141 Starlifter was one of the largest planes to use air station runways during the 1970s. The mammoth aircraft, a kind of flying hospital, was capable of long-range flights. In addition to the cockpit crew, it carried a medical team of two nurses and three medical technicians to care for the wounded that were aboard. After arriving at NASG, patients were transferred to special buses and ambulances for the trip to Great Lakes Naval Hospital.

The army became a tenant of the navy in 1976. They flew UH-1H "Huey" helicopters. The army eventually built its own hangar. Prior to that time, marine mechanics maintained army aircraft. The Army Reserve flight facility had formerly been located at Du Page County Airport. Army recruiting flight detachment followed in July 1977, after closure of Haley Army Airfield at Fort Sheridan, Illinois. The army was the last tenant to vacate NASG. About one month following closure of the air station, the 244th Aviation Brigade relocated operations to Fort Sheridan on October 17, 1995.

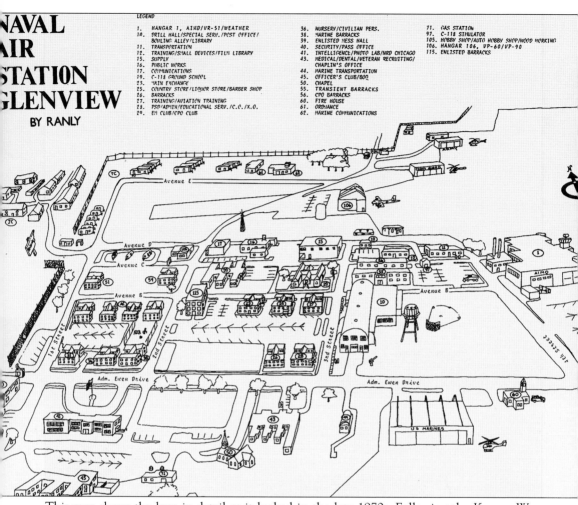

LEGEND

1. HANGAR 1, AIMD/VR-51/WEATHER
10. DRILL HALL/SPECIAL SERV./POST OFFICE/ BOWLING ALLEY/LIBRARY
11. TRANSPORTATION
12. TRAINING/SMALL DEVICES/FILM LIBRARY
15. SUPPLY
16. PUBLIC WORKS
17. COMMUNICATIONS
23. C-118 GROUND SCHOOL
23. MAIN EXCHANGE
25. COUNTRY STORE/LIQUOR STORE/BARBER SHOP
26. BARRACKS
27. TRAINING/AVIATION TRAINING
28. PSD/ADMIN/EDUCATIONAL SERV./C.O./X.O.
29. EM CLUB/CPO CLUB

36. NURSERY/CIVILIAN PERS.
38. MARINE BARRACKS
39. ENLISTED MESS HALL
40. SECURITY/PASS OFFICE
41. INTELLIGENCE/PHOTO LAB/NRD CHICAGO
43. MEDICAL/DENTAL/VETERAN RECRUITING/ CHAPLIN'S OFFICE
44. MARINE TRANSPORTATION
45. OFFICER'S CLUB/BOQ
50. CHAPEL
55. TRANSIENT BARRACKS
56. CPO BARRACKS
60. FIRE HOUSE
61. ORDNANCE
62. MARINE COMMUNICATIONS

71. GAS STATION
97. C-118 SIMULATOR
105. HOBBY SHOP/AUTO HOBBY SHOP/WOOD WORKING
106. HANGAR 106, VP-60/VP-90
115. ENLISTED BARRACKS

This map shows the base in detail as it looked in the late 1970s. Following the Korean War, more improvements were made to the infrastructure at the air base, with emphasis on safety for both pilots and residents of Glenview. Among these were barrier assemblies that were installed on both ends of runway 17/35 to facilitate quick stops by jets.

During the 26 years the Coast Guard was located at Glenview, crews flew search and rescue missions over southern Lake Michigan and other bodies of water in the area. Hundreds of boaters owed their lives to the Coast Guard, who flew at all hours and under hazardous conditions. The HH52A Sea Guard helicopters flown by the Coast Guard when they arrived at the base were later replaced by the HH65A Dolphin.

Patrol squadrons began to receive P-3 Orions as replacements for the old P-2 Neptunes in 1974. The P-3s were equipped with the latest submarine-tracking electronics.

Glenview Naval Air Base
Airplane cockpit(#2)

Pilot Richard Baker and engineer Mel Henschel perform a pre-flight check on a P-3 Orion.

This view inside a P-3 Orion shows radar consoles and other electronic systems that were used to track submarines.

When the old tower atop Hangar One became inadequate, a new five-story structure was built in 1979 near the east gate of the base. It provided a panoramic view of the runways.

The construction cost was $1.5 million; the tower contained the latest technology for aircraft control. Handling its final takeoffs and landings in 1994, the tower's lifespan was only 14 years.

First Lady Betty Ford, wife of Pres. Gerald Ford, is greeted by NASG's commander in 1977. Her husband had been stationed at the base nearly 40 years earlier.

Vice Pres. George H. W. Bush and Barbara Bush arrive at NASG aboard Air Force Two in 1988, some 44 years after Vice President Bush was briefly attached to the CQTU during World War II. The purpose of this visit was the rededication of the Admiral Gallery Memorial at NASG.

A ceremony to mark the 45th anniversary of NASG took place in 1982. Base commander Capt. S. B. Palmer is pictured at the podium. Vice Adm. Richard F. Whitehead, U.S. Navy, was guest of honor and keynote speaker. Station chaplain Lt. Comdr. F. B. Burchell looks on. World War II aircraft flew over, and modern planes were opened for inspection on the ground. The event was attended by 600 people, including citizens from Glenview and Northbrook.

On April 20, 1989, NASG squadron VP-90, Crew 3, was participating in a sea training exercise in the South China Sea. During the mission, the crew encountered a Soviet "charlie" class nuclear submarine. The submarine surfaced, allowing Crew 3 to obtain a great deal of intelligence. The crew was recognized for their work. This print, titled *Sorry Charlie*, was commissioned to commemorate the event. (Courtesy of VP-90.)

Base commander Capt. W. D. Vandivort greets army general Colin Powell upon arrival his at NASG around 1990. General Powell served as chairman of the Joint Chiefs of Staff and later was secretary of state during the first administration of Pres. George W. Bush.

Seven

OVER AND OUT

The last military action for personnel from NASG was Operation Desert Storm. The conflict was launched in mid-January 1991; some 250 navy and marine reservists were called up. VMGR-234 (Marine Aerial Refueling Transport Squadron) was activated on January 9, 1991. HML-776 (Marine Light Helicopter Squadron) was mobilized on January 22, 1991. MACG-48 (Marine Air Control Group) served in the Philippines for 10 months to provide support for the U.S. Embassy in Manila and for support of Cubi Point Naval Air Station.

The beginning of the end for NASG came with the establishment of the Base Realignment and Closure (BRAC) Commission in 1988. Glenview was on the final closure list published in early 1993. By autumn of that year, plans were underway to discontinue operations and close the facility.

In due course, the Department of Defense awarded ownership of the former NASG site to the Village of Glenview. A small portion of the property has been retained for housing for navy personnel; another has been set aside to preserve a vestige of native prairie. Runways were demolished to make way for streets and roads. Instead of barracks, there are upscale homes and retirement communities. Mess halls have given way to restaurants and base exchanges to retail stores. Portions of Hangar One, including the control tower, were incorporated into new commercial construction. Gallery Park (named in honor of Adm. Dan Gallery) provides space and a lake for leisure activities. Navy Park is situated so that the old Hangar One tower overlooks bronze statues depicting an aviator, sailor, and "yellow shirt" (carrier deck crewman). Schram Memorial Chapel was moved, restored, and, now owned by the Glenview Park District, is called the Schram Memorial Museum. The Glenview Area Historical Society and Glenview Hangar One Foundation have worked together to preserve the base's history. The foundation maintains a museum adjacent to the former base property, and the Glenview Area Historical Society holds extensive archives of pictures, as well as written material related to the air station's history.

NASG is gone, but its place in American history is not forgotten.

It almost seemed an omen of things to come when, early in the morning on March 13, 1992, fire broke out in the Drill Hall. Ghostly remnants of a staircase to nowhere marked the site. The base water tower is silhouetted in the background.

Firefighters from the base, the village of Glenview, and surrounding suburbs tried, to no avail, to save the 50-year-old Building 10, also known as Drill Hall. Despite their best efforts, the building burned to the ground. One of the most venerable structures on the base was gone. Drill Hall had served as a training and recreation facility for 50 years. Its spaces provided both military and civilians with a venue for events; the loss was keenly felt.

One of the last aerials, taken around 1993, shows the look of NASG at the time the decision was made to close the base. Damage to the ground from the Drill Hall fire is still visible. (Courtesy of Lt. Comdr. Harley Hett.)

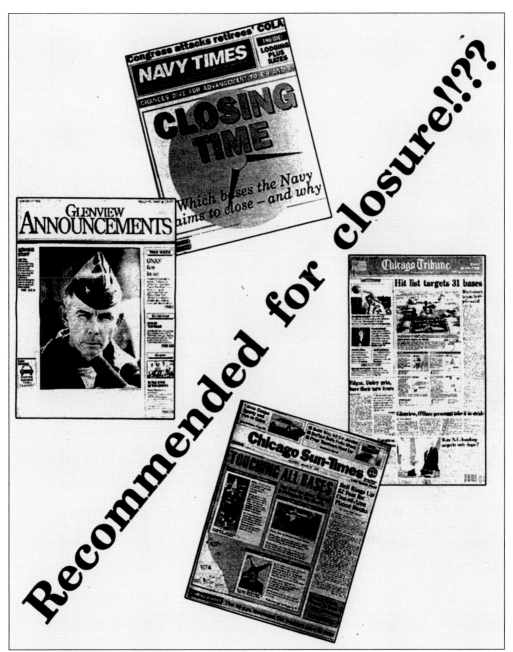

Chicago-area newspaper headlines announced the fact that NASG appeared on the BRAC list. In a photograph on the left side of the collage, NASG commander Capt. P. W. Kinneberg appears perplexed—would the base be closed or not? NASG had survived the 1989 and 1991 rounds of closures but was not so fortunate in 1993.

Photo by LT Kurt Po

Patrol Squadrons 60 and 90 fly into sunse

When the decision was announced that NASG would indeed close, squadrons were disbanded, and aircraft and personnel were relocated. Patrol Squadrons VP-60 and VP-90 were decommissioned in February 1994. Both squadrons had served with distinction for nearly 25 years and received numerous awards during their tenure at NASG. The P-3 Orions that they flew were reassigned, as the number of aircraft based at NASG continued to dwindle.

In Hangar One's ready room, NASG commanding officer Capt. James Schultz (right) confers with fellow officers on February 26, 1995, as the last flights prepare to take off from base runways. A large crowd, both military and civilian, stood by as three P-3 Orions revved their engines and prepared for departure.

Last Aircraft To Depart N.A.S. Glenview 26 Feb 1995

Three P-3 Orion crews were the last to drill at NASG. Here a crew member waves to those assembled to witness the historic event. One by one, the P-3 Orions rolled down the runway, lifted off, and disappeared into the overcast sky. Coming around, the last plane made a final pass over the runway, dipped its wings in salute, and flew away. A ground crew moved out and painted each end of the runway with a yellow X, denoting a closed airfield. Flight operations at NASG were finished forever.

NAVAL AIR STATION GLENVIEW

CLOSURE CEREMONY
SEPTEMBER 9, 1995

September 9, 1995, was the date set for the decommissioning of NASG. On this bright, warm day, more than 1,700 people—civilian as well as military—attended the impressive ceremony. The program for base closure ceremonies heralded the end of an era. The Seabee Unit based at Glenview had removed and finished small pieces of concrete from the runways; these souvenirs were given to everyone who attended the ceremony.

There were speeches by Capt. James Schultz, NASG commanding officer, and Rear Adm. J. D. Olson. The American flag flying above Hangar One was lowered for the last time; the name was now former Naval Air Station Glenview. Over the years, public feeling regarding the base ranged from complete support to uneasy coexistence. However, it is likely that this crowd felt anything but regret at the end of the day.

The base stood virtually abandoned for a few years. Hangar One remained the only building determined to be eligible for nomination to the National Register of Historic Places.

The State of Illinois
and
United States Department of the Interior
National Park Service
announce the listing of

Hangar 1, Naval Air Station, Glenview

in the

National Register of Historic Places

November 12, 1998
Springfield, Illinois

William L. Wheeler
State Historic
Preservation Officer

Members of the Glenview Area Historical Society and the Glenview Hangar One Foundation submitted the National Register of Historic Places nomination. The effort was successful, with Hangar One officially named to the National Register of Historic Places on Veteran's Day, November 11, 1998.

As work began to redevelop the former air station, large portions of Hangar One were demolished to make way for new construction. The tower and portions of the east facade, along with the north and south pods, or fishbowls, were incorporated into new buildings. This view looking east reveals the extent of demolition to the midsection of the hangar under the tower.

Portions of the south hangar wall (facing east) remain, as does the shell of the south fishbowl. New construction took place around these portions, leaving the facades in place. It is estimated that about 80 percent of the original hangar was torn down. According to professional historic preservationists, this is why Hangar One no longer qualifies as a National Register of Historic Places property.

A perspective of Glen Town Center shows the old Hangar One tower rising above Navy Park. It is now surrounded by retail stores and restaurants. Displayed in the park are three statues depicting a sailor, a pilot, and a deck crewman. Several hundred commemorative bricks, most in honor of veterans, have been set into the grounds of the small park. (Courtesy of Bill Dawson.)

ACROSS AMERICA, PEOPLE ARE DISCOVERING SOMETHING WONDERFUL. *THEIR HERITAGE.*

Arcadia Publishing is the leading local history publisher in the United States. With more than 3,000 titles in print and hundreds of new titles released every year, Arcadia has extensive specialized experience chronicling the history of communities and celebrating America's hidden stories, bringing to life the people, places, and events from the past. To discover the history of other communities across the nation, please visit:

www.arcadiapublishing.com

Customized search tools allow you to find regional history books about the town where you grew up, the cities where your friends and family live, the town where your parents met, or even that retirement spot you've been dreaming about.